PUPPIES

CHECKERBOARD PRESS NEW YORK

Come on, sleepyheads,
wake up!
We'll ask big Patches
to take us for a romp
in the country.

We're tired of sitting.
Will someone play ball
with us?

Has everyone forgotten us? How about lunch? We're hungry.

We may look small now, but just wait — we'll be grown-up dogs one day.
Woof, woof, woof!

Peanuts is going to get in trouble if she doesn't stop chewing that brush. But it sure looks like fun!

That basket
is much too crowded.
It's more comfortable
in the hay.

Please, don't stare.
It's not polite.

When we get a little bigger, we'll have nice collars, too — just like our friends.

We're not lost.
We're playing
hide-and-seek.
Come and find us.

My, my, what's there?
A butterfly?
A caterpillar?
Let's go find it.

We are all
good puppies.
Will you take
us home?

CHECKERBOARD PRESS Colophon are trademarks of Macmillan, Inc.